FLORA OF TROPICAL EAST AFRICA

BLECHNACEAE

B.S. PARRIS[1]

Usually terrestrial ferns, sometimes lithophytic, epiphytic or climbing. Rhizomes erect to long-creeping, scaly, scales not clathrate. Stipes often scaly, particularly at base. Lamina simple (not in our area), pinnatifid to pinnate, rarely bipinnate, fertile fronds with much reduced lamina or fertile and sterile fronds similar (not in our area); young fronds sometimes reddish, sori elongate, in one or more rows on each side of pinna mid-vein, indusia (when present) opening towards pinna mid-vein.

A cosmopolitan family of eight genera and about 250 species.

Rhizome erect or creeping, rarely climbing on trees; fertile fronds pinnate; indusium present, linear .... 1. **Blechnum**
Rhizome creeping on ground or climbing on trees; fertile fronds sometimes bipinnate; indusium absent .... 2. **Stenochlaena**

1. BLECHNUM

L., Sp. Pl.: 1077 (1753); Copeland, Gen. Fil.: 155 (1947); Kramer et al. in Kubitzki et al., Fam. Gen. Vasc. Pl. 1: 63 (1990)

Terrestrial, epiphytic, lithophytic or climbing; rhizomes erect, short- or long-creeping, radial; stipes in whorls, not articulated to rhizome; rhizomes covered with narrowly lanceolate pale to dark brown or blackish glabrous subglossy to glossy scales, sometimes with paler margin; stipes with scales like those of rhizome when young, particularly in basal part. Fronds pinnate or pinnatifid (rarely bipinnatifid, not in our area); fertile fronds pinnate, with much reduced lamina (fertile and sterile fronds sometimes similar, not in our area); young fronds often reddish; lateral veins free in sterile fronds; sori linear, one (or one row of sori) on each side of pinna mid-vein, indusia opening towards pinna mid-vein.

A cosmopolitan genus of approximately 150–200 species, widely distributed in tropical montane and south temperate areas.

1. Sterile pinnae completely adnate at base in middle part of frond 2
   Sterile pinnae free or adnate only on basal side in middle part of frond 3
2. Lowest sterile pinnae reduced to auricles, margin of sterile pinnae entire (fig. 1) .... 1. *B. attenuatum*
   Lowest sterile pinnae not reduced to auricles, margin of sterile pinnae minutely crenate-serrate .... 2. *B. ivohibense*

[1] Fern Research Foundation, 21 James Kemp Place, Kerikeri, Bay of Islands, New Zealand

3. Sterile pinnae cuneate, free or adnate on basal side in middle pinnae (fig. 1) .... 3. *B. tabulare*
   Sterile pinnae auriculate, free on both sides in middle pinnae 4
4. Pinnae apiculate at apex, with minutely serrate margins; indusium erose to lacerate .... 4. *B. australe*
   Pinnae acute to attenuate at apex, with entire margins; indusium entire .... 5. *B. punctulatum*

1. **Blechnum attenuatum** *(Sw.) Mett.*, Fil. Hort. Bot. Lips.: 64, t. 3, fig. 1–6 (1856); Tardieu-Blot in Mém. I.F.A.N. 28: 201, t. 39, fig. 7 (1953); F.W.T.A. ed. 2, Suppl.: 74 (1959); Tardieu-Blot, Fl. Madag., Polypod. 2: 8, fig. 2, 5–7s (1960) & Fl. Cameroun 3: 295, t. 34, fig. 7 (1964); Schelpe in F.Z. Pterid.: 235 (1970) as var. *attenuatum*; Burrows, S. Afr. Ferns & Fern Allies: 326, fig. 330, t. 54, 4 (1990) as var. *attenuatum*; Iversen in Symb. Bot. Upsal. 29(3): 156 (1991); Johns, Pteridophytes T.E.A.: 98 (1991). Type: Mauritius, *Groendal* s.n. (P, holo.)

Rhizomes 6–40 mm in diameter including scales, ± erect to long-creeping; scales 12–20 × 1–3 mm, linear-lanceolate, pale to dark chestnut-brown, sometimes with medium brown to blackish central stripe, sometimes crisped at apex. Stipes 5–28 × 0.2–0.6 cm. Fronds pinnate or very deeply pinnately divided, fertile and sterile laminae similar in size, 23–120(–200?) × 5–32 cm, narrowly elliptic in outline, middle pinnae adnate at base, lowest pinnae of sterile fronds reduced to auricles, middle pinnae acuminate at apex, decurrent at base, longest sterile pinnae 40–210 × 8–19 mm, very narrowly triangular-oblong to linear, margin entire, sometimes falcate, longest fertile pinnae linear, 30–180 × 2–4 mm. Sori extending along most of fertile pinnae, unbroken, indusium entire to slightly erose, 0.4–0.6 mm wide. Fig. 1/1, 5 (p. 3).

UGANDA. Toro District: Ruwenzori, R. Nyamugasani, 22 Aug. 1952, *Osmaston* 2191! & Ruwenzori, Wimi, *Scott Elliot* 7882!; Kigezi District: Impenetrable Forest, Sept. 1936, *Eggeling* 3252!

KENYA. Meru District: volcanic cone Kirui, E slope of Mt Kenya, 13 April 1969, *Faden et al.* 69/497!; Kericho District: SW Mau Forest, along the Kiptiget [Chepkoisi] R. ±16 km SSE of Kericho, 12 June 1972, *Faden et al.* 72/345!; Teita District: Kasigau, Rukanga route, 16 Nov. 1994, *Luke & Luke* 4124!

TANZANIA. Arusha District: Mt Meru, Engare Nanyuki R., 3 March 1971, *Richards* 26831!; Lushoto District: W Usambaras, Mbalamu, Shagai forest, 15 March 1893, *Holst* 2479!; Morogoro District: Ngata River above Huala Falls, 24 Aug. 1951, *Greenway & Eggeling* 8864!

DISTR. **U** 2; **K** 3–5, 7; **T** 2, 3, 6–8; Cameroon, Bioko, São Tomé, Zambia, Malawi, Zimbabwe; Madagascar, Comoro Islands, Réunion, Mauritius

HAB. Pendulous epiphyte on trunks of trees and of tree ferns, particularly *Cyathea manniana*, sometimes of *C. dregei*, or terrestrial or lithophytic, in moist montane ± evergreen forest, or in scrubby vegetation, on rocky banks of streams and rivers, often in the spray zone of waterfalls; 1500–3000 m

SYN. *Onoclea attenuata* Sw. in Schrad., J. Bot. 1800, 2: 73 (1801)

NOTE. Schelpe (J. Linn. Soc. Bot. 53: 493, 1952) cites the type of *B. attenuatum* (Sw.) Mett. var. *holstii* (Hieron.) Schelpe as being in K. Material of the type number, *Holst* 2479, at K and B belongs to a small form of *B. attenuatum*, but other material of this collection at B and at P is referable to *B. ivohibense*. One sheet at B bears the label 'Blechnum Holstii Hieron. steht dem B. polypodioides nahe. Det. Georg Hieronymus'. As this is the only sheet seen that has been annotated by Hieron. it is selected as the lectotype of var. *holstii* (see under *B. ivohibense*).

Blechnum attenuatum is very variable in our area in habit, thickness of rhizome and colour of rhizome scales. Fronds of juvenile plants have a long undivided lamina apex. *Blechnum giganteum* (Kaulf.) Schltdl. is a closely related Southern African species to which some East African material has been referred (Johns, Pterid. T.E.A.: 98, 1991). It has been distinguished from *B. attenuatum* by its larger size and terrestrial or lithophytic, rather than epiphytic (rarely lithophytic), habit, and rhizome 20–34 mm in diameter, rather than less than 20 mm in diameter. (Burrows, S. Afr. Ferns & Fern Allies: 326, 1990; Jacobsen, Ferns S. Afr.: 460, 1983; F.S.A. Pteridophyta, 269, 1986). Chambers (annotations on various herbarium sheets

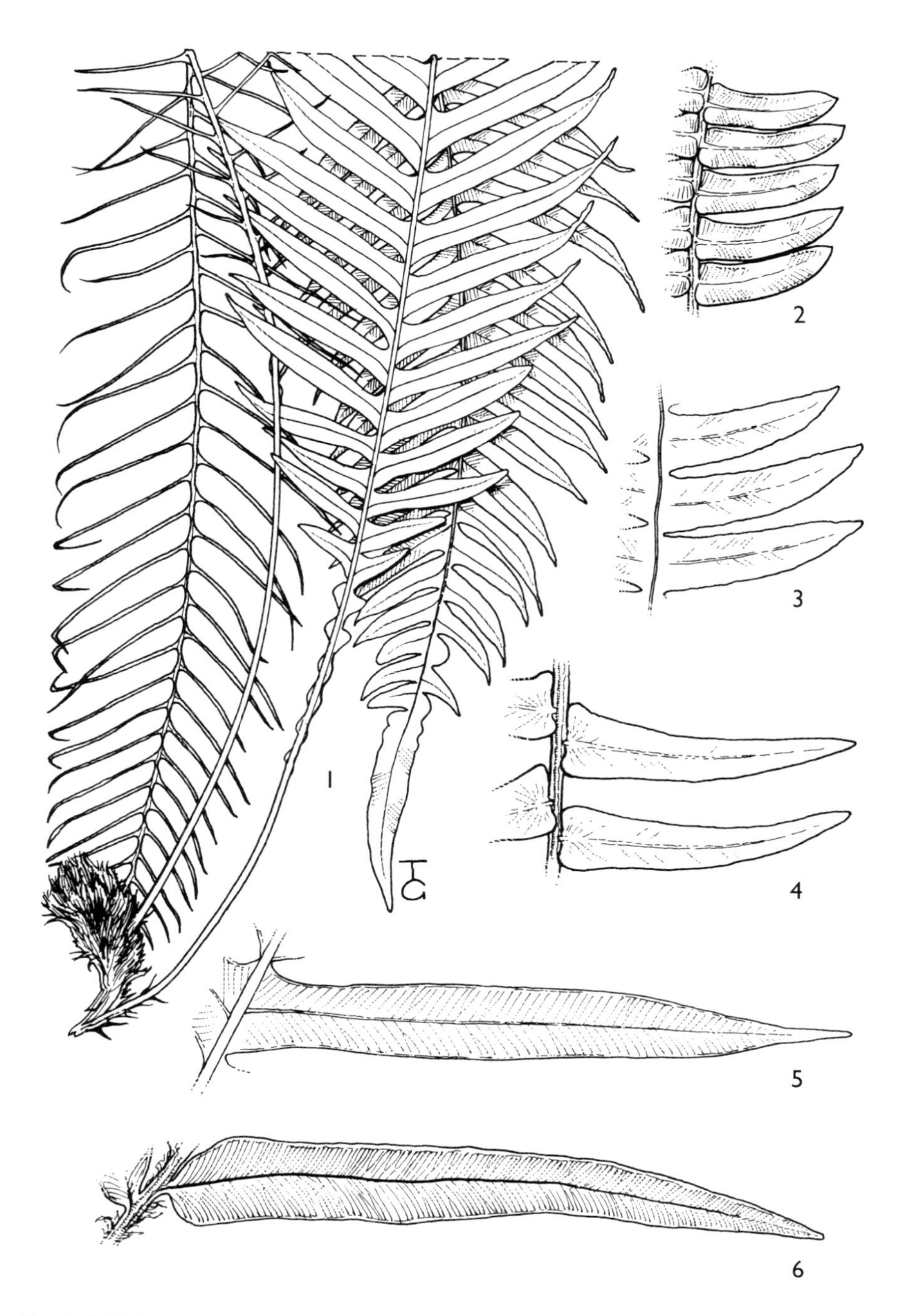

FIG. 1. *BLECHNUM ATTENUATUM* — **1**, habit, × ⅖; *B. AUSTRALE* — **2**, pinnae, × 1; *B. IVOHIBENSE* — **3**, pinnae, × 1; *B. PUNCTULATUM* — **4**, pinnae, × 1; *B. ATTENUATUM* — **5**, pinna, × 1; *B. TABULARE* — **6**, pinna, × 1. 1, 5 from *Vessey-Fitzgerald* 4827; 2 from *Gilbert* 6315; 3 from *Thomas* 3884; 4 from *Thulin & Mhoro* 2811; 6 from *Atlen* 3740. Drawn by Tim Galloway.

at K), seems to regard bicolorous scales as indicative of *B. attenuatum* and concolorous scales as characteristic of *B. giganteum*, but I have seen concolorous scales on plants of epiphytic habit with slender rhizomes and small fronds that are otherwise clearly referable to *B. attenuatum*. Plants may also have both concolorous and bicolorous scales. It is impossible to distinguish two species in East African material using the characters mentioned above because of the degree of intergradation between them. A form of *B. attenuatum* with very small fronds widely spaced on a slender creeping rhizome was distinguished by Schelpe (J. L.S. Bot. 53: 493, 1952) as var. *holstii*, but this name is a synonym of *B. ivohibense* (q. v.). This form merits further study in the context of variation within and between *B. attenuatum* and *B. giganteum*.

2. **Blechnum ivohibense** *C.Chr.* in Arch. Bot. (Caen) 2: Bull. Mens.: 211 (1928); Schelpe in J.L.S. Bot. 53: 493 (1952) & in Brit. Fern Gaz. 9(8): 349 (1967) & in F.Z. Pterid.: 237, t. 68 (1970); Jacobsen, Ferns S. Afr.: 462, t. 348 (1983); Burrows, S. Afr. Ferns & Fern Allies: 330, fig. 332, t. 55, 1 (1990); Iversen in Symb. Bot. Upsal. 29(3): 156 (1991); Johns, Pterid. T.E.A.: 98, f. 23.1 (1991). Type: Madagascar, Pic d'Ivohibe, Bara, *Humbert* 3300 (BM, holo.)

Rhizomes 8–14 mm in diameter including scales, erect to short-creeping; scales 5–7 × 1 mm, narrowly lanceolate, dark brown to black. Stipes 4–28 × 0.1–0.3 cm, longer in fertile than in sterile fronds. Fronds ± erect, pinnate or deeply pinnatifid, lamina narrowly elliptic to narrowly lanceolate in outline, lowest pinnae not reduced to auricles; fertile laminae 160–280 × 20–45 mm, longest pinnae 11–33 × 1–2 mm, linear, acuminate at apex, sessile on acroscopic margin, sessile to decurrent on basiscopic margin at base; sterile laminae 12–41 × 2.2–10 cm, longest pinnae 1.4–5.5 × 0.5–1.2 cm, narrowly triangular-oblong, sometimes falcate, obtuse to acute at apex, sessile at base, margin minutely crenulate-dentate. Sori extending along most of fertile pinnae, unbroken, indusium entire, 0.1–0.3 mm wide. Fig. 1/3 (p. 3).

KENYA. Meru District: below bridge along road west of volcanic cone Kirui on E slope of Mt Kenya, 13 April 1969, *Faden et al.* 69/495!; Marimba, Ithangune forest, road along base of volcanic cone Kirui, 10 miles from Nkubu, 22 June 1969, *Faden* 69/773 *et al.*!; Marimba Forest, 2nd bridge on road to Volcanic Cone Kirui, on slopes of Ithangune, NE side of Mt Kenya, 28 Feb. 1970, *Faden & Evans* 70/77!

TANZANIA. Kilosa District: Ukaguru Mts N of Kilosa, Mandege Forest Station, 29 July 1972, *Pócs & Mabberley* 6737/F!; Morogoro District: Uluguru Mts, SE ridge of Bondwa, 26 Sept. 1970, *Faden* 70/643 *et al.*!; Iringa District: Mwanihana Forest Reserve above Sanje village, 10 Oct. 1984, *D.W. Thomas* 3884!

DISTR. **K** 4; **T** 2, 3, 6–8; Mozambique, Zimbabwe; Madagascar

HAB. In dense forest and mist forest, in narrow gorges, under bridges and in caves, on soil banks and shaded vertical rock walls; 1500–2350 m

SYN. *B. polypodioides* (Sw.) Kuhn var. *holstii* Hieron. in Engler, P.O.A. C: 81 (1895); F.D.O.-A 1: 82 (1929) & Descript. (Anhang): 9, Fig. 5–8 (1929). Type: Tanzania, W Usambaras, Mbarama, *Holst* 2479, pro parte (lecto. B!, chosen here; P!, isolecto.)

B. umbrosum Peter in Repert. Spec. Nov. Regni Veg. Beih. 40, 2: 9, t. 3, fig. 5–8 (1929). Type: Tanzania, W Usambaras, Kisimba above Mazumbai, April 1916, *Peter* 16489 (B!, lecto. chosen here; BM!, BR!, K!, P photo seen in BM!, isolecto.)

B. attenuatum (Sw.) Mett. var. *holstii* (Hieron.) Schelpe in J.L.S. Bot. 53: 493 (1952)

NOTE. Schelpe (J.L.S. Bot. 53: 493, 1952) cites the type of *B. attenuatum* (Sw.) Mett. var. *holstii* (Hieron.) Schelpe as at K. Material of *Holst* 2479 at K and B belongs to a small form of *B. attenuatum*, but other material of this collection at B and at P is referable to *B. ivohibense*. One sheet at B, B 031479, bears the label 'Blechnum Holstii Hieron. steht dem B. polypodioides nahe. Det. Georg Hieronymus', with 'det. Georg Hieronymus' printed, the rest in Hieronymus' handwriting, and another label with 'Museum botanicum Berolinense' printed and 'Blechnum polypodioides var. Holstii Hieron.' in Hieronymus' handwriting. As this is the only sheet seen that has been annotated by Hieron. it is selected here as lectotype. The sheet in P (00113566) has a label written in a different hand 'Blechnum polypodioides (Sw.) Kuhn determ. Hieronymus'. *Blechnum bakeri* C. Chr. from Madagascar is rather similar but has fewer pinnae and its laminae are setose abaxially.

3. **Blechnum tabulare** *(Thunb.) Kuhn*, Fil. Afr. 94 (1868); Sim, Ferns S. Afr. ed. 2: 187, t. 83 (1915); F.D.O.-A 1: 82 (1929); F.W.T.A. ed. 2, Suppl.: 74 (1959); Tardieu-Blot, Fl. Madag., Polypod. 2: 3 (1960) & Fl. Cameroun 3: 294 (1964); Schelpe in F.Z. Pterid.: 237 (1970); Roux in Cape Penin. Ferns: 60 (1979); Schelpe in C.F.A. Pterid.: 184 (1977); Jacobsen, Ferns S. Afr.: 463, t. 42, 349 (1983); Schelpe & Anthony, F.S.A. Pterid.: 271, fig. 93 (1986); Burrows, S. Afr. Ferns & Fern Allies: 331, fig. 334, t. 55, 3 (1990); Johns, Pterid. T.E.A.: 98 (1991). Type: South Africa, Cape Province, Cape Peninsula, summit of Table Mountain, *Thunberg* s.n. (UPS, holo.; S, iso., BOL, photo.).

Rhizomes 3–15 cm in diameter including scales, erect, to 100 cm tall; scales 25–35 × 1–2 mm, linear-lanceolate, dark brown to blackish with paler margins. Stipes 4–52 × 0.3–0.6 cm, longer in fertile than in sterile fronds. Fronds erect, pinnate, fertile and sterile laminae similar in size, 40–120 × 3–25 cm, narrowly elliptic in outline, middle pinnae bluntly acute to acuminate at apex, middle sterile pinnae free or adnate on basiscopic margin, middle fertile pinnae free, lower pinnae reduced, sometimes to auricles, longest sterile pinnae 7–18 × 0.8–2.2 cm, very narrowly elliptic to cuneate base, margin entire, often inrolled, longest fertile pinnae linear, 4–15 × 0.2–0.6 cm. Sori extending along most of fertile pinnae, unbroken, indusium lacerate, 0.5–2 mm wide. Fig. 1/6 (p. 3).

UGANDA. Kigezi District: Bufumbira, Nyamagana, June 1951, *Purseglove* 3693! & Bufindi, Lake Bunyoni, Dec. 1938, *Chandler & Hancock* 2533! & near Kabale, *Molesworth-Allen* 3740!

KENYA. Naivasha District: Nyandarua, Sasamua Dam, just below outlet, 24 Jan. 1971, *Faden* 71/72 *et al.*!; Kiambu District: Sasamua Dam, near South Kinangop, 3 Aug. 1981, *Gilbert* 6337!; Meru District: volcanic cone Kirui, NE side of Mt Kenya, 28 Feb. 1970, *Faden* 70/90 *& Evans*!

TANZANIA. Iringa District: Mufindi, Kibwele Estate, Luisenga stream, just below dam at Mufindi Rod and Gun Club fishing lodge, 30 Jan. 1989, *Gereau & Lovett* 2984! & Kigogo Forest Reserve, 11 km by road below (South) of the Mufindi – James Corner Road, 20 Aug. 1971, *Perdue & Kibuwa* 11185! & Dabaga, near Kidabaga village, 26 Aug. 1984, *Thomas* 3593!

DISTR. **U** 2; **K** 3, 4; **T** 2–4, 6–8; Nigeria, Cameroon, Congo (Kinshasa), Angola, Zambia, Malawi, Mozambique, Zimbabwe, South Africa; Madagascar

HAB. Terrestrial in *Juniperus procera-Arundinaria alpina* mixed bamboo forest, disturbed forest, wet grassland, peat bogs & swamps, in shade amongst rocks below waterfall and on rocks in dry situations; 1600–2600 m

SYN. *Pteris tabularis* Thunb. in Prod. Pl. Cap.: 171 (1800)

NOTE. Used as an antibiotic (*Lovett & Thomas* fide 2381)

4. **Blechnum australe** *L.*, Mant. Pl.: 130 (1767); Sim, Ferns S. Afr. ed. 2: 188, t. 84 (1915); Tardieu-Blot, Fl. Madag., Polypod. 2: 4, fig. 1, 3 (1960); Launert in Prodr. Fl. SW-Afr. 8: 1 (1969); Schelpe in F.Z. Pterid.: 240 (1970); Roux, Cape Penin. Ferns: 61 (1979); Jacobsen, Ferns S. Afr.: 469, t. 352 (1983); Schelpe & Anthony, F.S.A. Pterid.: 273, fig. 94, 1 (1986); Burrows, S. Afr. Ferns & Fern Allies: 335, fig. 336, t. 56, 2 (1990); Johns, Pterid. T.E.A.: 98 (1991). Type: South Africa, Cape Province, Cape Peninsula (LINN 1247/3, holo.).

Rhizomes 3–20 mm in diameter including scales, erect, sometimes stoloniferous; scales 2–10 × 0.2–2.0 mm, narrowly lanceolate, dark brown with paler margin. Stipes 20–200 × 0.5–2.0 mm. Fronds erect, pinnate, fertile and sterile laminae similar in size, 6–58 × 1–6 cm, narrowly elliptic in outline, lower pinnae reduced to auricles, pinnae apiculate at apex, auriculate at base, margin minutely serrate, longest sterile pinnae 7–30 × 3–8 mm, very narrowly triangular-oblong, longest fertile pinnae 5–35 × 1–4 mm, linear beyond dilated base. Sori extending along most of fertile pinnae, unbroken, indusium erose or lacerate, 0.3–0.8 mm wide. Fig. 1/2. (p. 3).

KENYA. Naivasha District: Nyandarua, Sasamua Dam, 24 Jan. 1971, *Faden et al.* 71/71! & Sasamua Dam, S Kinangop, 13 Sept. 1970, *Cameron* 18! & 3 Aug. 1981, *Gilbert* 6315!

TANZANIA. Arusha District: Mt Meru East Slope, Jekukuma R., 21 March 1966, *Greenway & Kanuri* 12462!; Moshi District: S of Kilimanjaro, 17 Sept. 1910, *Daubenberger* s.n.!; Kigoma District: Mahali Mts, Sisaga, 1.5 km S of summit, 28 Aug. 1958, *Newbould & Jefford* 1839!
DISTR. **K** 3–5; **T** 2, 4, 7; Zimbabwe, South Africa; Madagascar, Tristan da Cunha & Gough Islands
HAB. Terrestrial on steep slopes in *Juniperus procera–Arundinaria alpina* mixed bamboo forest, in grassy herbage, in rock crevices, in moist conditions in sun or shade; 1500–2500 m

5. **Blechnum punctulatum** *Sw.* in Schrad., J. Bot. 1800, 2: 74 (1801); Tardieu-Blot, Fl. Madag., Polypod. 2: 6, fig. 1, 1–2 (1960); Schelpe in F.Z. Pterid.: 239 (1970); Roux, Cape Penin. Ferns: 60 (1979); Jacobsen, Ferns S. Afr.: 466, t. 351a (1983); Schelpe & Anthony, F.S.A. Pterid.: 273, fig. 95 (1986); Burrows, S. Afr. Ferns & Fern Allies: 332, fig. 335, t. 55, 4 (1990); Iversen in Symb. Bot. Upsal. 29(3): 156 (1991); Johns, Pterid. T.E.A.: 98 (1991). Type: South Africa, Cape Province, Cape Peninsula, *Thunberg* s.n. (S, holo., BOL photo.)

Rhizomes 1.5–2.5 cm in diameter including scales, erect or shortly creeping; scales 10–15 × 1–2 mm, narrowly lanceolate, dark brown, sometimes with paler margin. Stipes 3.5–9 × 0.2–0.5 cm. Fronds erect, pinnate, fertile and sterile laminae similar in size, 55–110 × 5–15 cm, narrowly elliptic in outline, middle pinnae free, lower pinnae reduced to auricles, pinnae acuminate at apex, auriculate at base, margin entire, longest sterile pinnae 3–8 × 0.7–1.2 cm, very narrowly triangular-oblong, longest fertile pinnae 5–8.5 × 0.2–0.4 mm, linear beyond dilated base. Sori extending along most of fertile pinnae, unbroken or rarely with 1–5 separate pairs of sori at base of pinna, indusium entire, 0.4–0.7 mm wide. Fig. 1/4 (p. 3).

TANZANIA. Moshi District: Kilimanjaro, S slope between Umbwe and Weru Weru Rivers, 31 Aug. 1932, *Greenway* 3198!; Kilosa District: Ukaguru Mountains N of Kilosa, Mamiwa Range, near Mnyera peak, 30 July 1972, *Pócs* 6740/A; Morogoro District: Uluguru Mts, NW slopes of Lupanga, 14 Feb. 1970, *Pócs* 6125/P!
DISTR. **T** 2, 3, 6; Malawi, Zimbabwe, South Africa; Madagascar
HAB. In wet montane forest, forming very dense matted ground cover or very rarely epiphytic; 1300–2300 m

NOTE. *B. punctulatum* var. *atherstonei* (Pappe & Rawson) Sim of Southern Africa has the sori on the lower pinnae breaking up towards the rachis into separate sori (F.S.A. Pteridophyta, 275, 1986). Occasionally a few pairs of separate sori can be seen in East African material, but the character is very inconsistent, both on pinna of a single frond, and between fronds on the same plant. In our area plants with this soral arrangement do not merit separation as var. *atherstonei*.

STENOCHLAENA[2]

J. Sm. in J. Bot. Gött. 4: 149 (1941) & in Hook., London Journ. Bot. 3: 401 (1841)

Large ferns with rhizomes creeping along the ground but eventually becoming scandent epiphytes; rhizome-scales sparse. Fronds remote, dimorphic, pinnate or 2-pinnate. Sterile pinnae articulate with basal glands, firmly membranous or chartaceous with sharply cartilaginous serrate margins. Fertile pinnae pinnate or bipinnate, linear or divided into linear segments, almost entirely covered with sporangia below; paraphyses absent.

A small genus of the Old World tropics and sub-tropics which has been referred to the Polypodiaceae (sens. strict.) by Alston (1959). Schelpe agreed with Copeland (1947) that this genus is better referred to the Blechnaceae on the venation and spore characters and this view is followed here.

[2] By B. Verdcourt

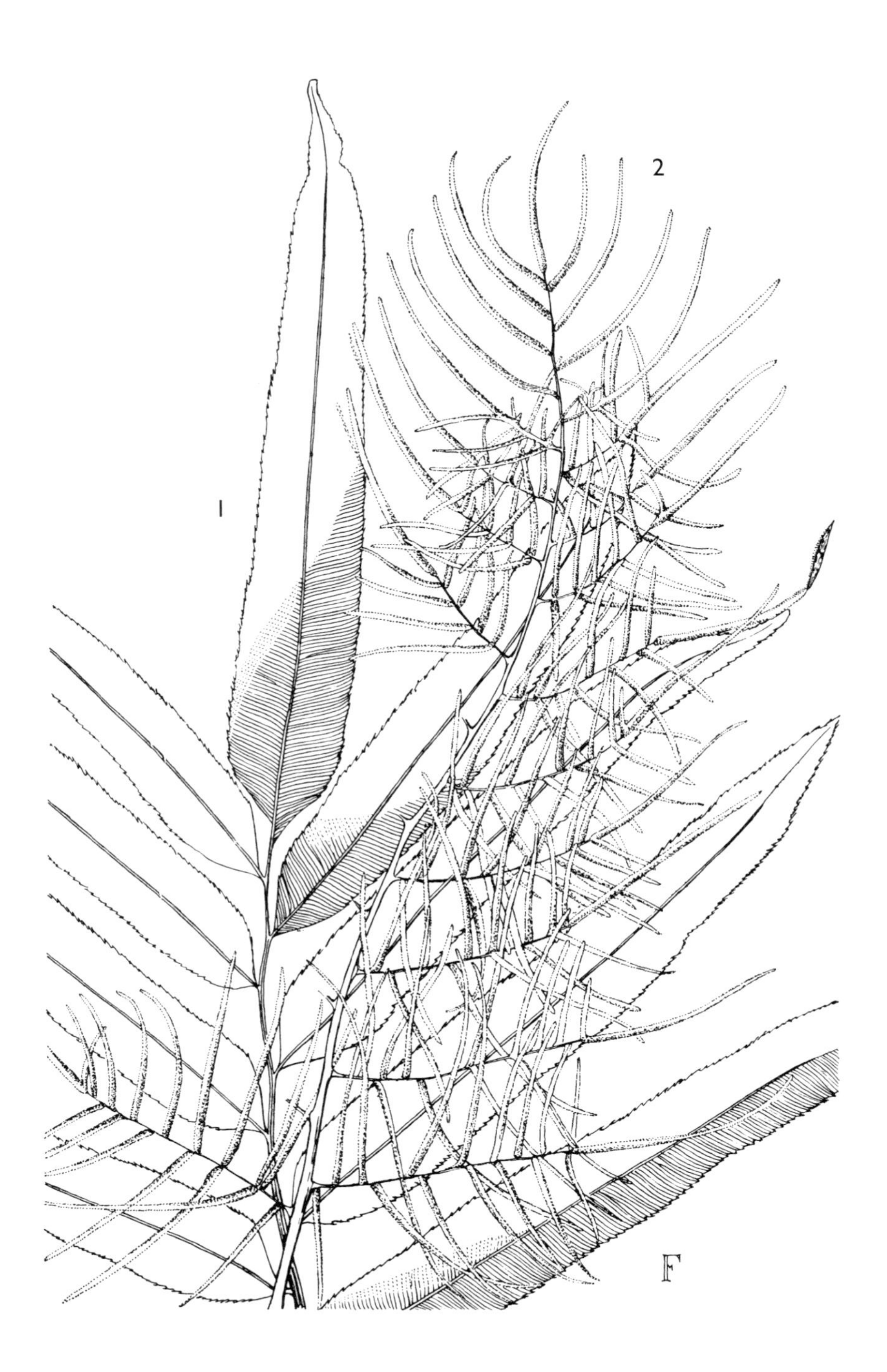

FIG. 2. *STENOCHLAENA TENUIFOLIA* — **1**, apical part of sterile frond, × ²⁄₃; **2**, apical part of fertile frond, × ²⁄₃. 1 from *Junod* 2; 2 from *Boivin* s.n. (Madagascar). From F.Z.

Stenochlaena tenuifolia *(Desv.) Moore* in Gard. Chron. 1856: 193 (1856); V.E. 2: 34 (1908); Sim, Ferns S. Afr. ed. 2: 192, t. 85, 86 (1915); F.D.O.-A 1: 83 (1929); U.O.P.Z.: 260 (1949); Tardieu-Blot, Fl. Madag. 5(1): 110, fig. 16/1–3 (1958); Schelpe in F.Z., Pterid.: 240, t. 69 (1970); Holttum in Amer. Fern J. 61: 120 (1971); Schelpe & Diniz, Fl. Moçamb. Pterid.: 252 (1979); Jacobsen, Ferns S. Afr. 471, t. 22, fig. 353, a, b (1983), Schelpe & Anthony, F.S.A., Pterid.: 278, map 241 (1986); Burrows, S. Afr. Ferns: 336, t. 56/3, fig. 82/337–337c, map (1990); Verdc. in K.B. 47: 128 (1992). Type: Madagascar, ? *Commerson* s.n. (P, holo., BOL, photo.; BM, iso.)

Rhizome up to 20 m long and 1–1.5 cm in diameter, creeping along the ground or ascending trees, sparsely clothed with dark-brown subulate rhizome-scales up to 5 mm long, becoming glabrous with age. Sterile fronds erect 1.4–4 m long, pinnate, fertile fronds erect, 2 m long, bi-pinnate or less often pinnate, both oblong in outline, widely spaced. Stipe pale brown, 30–50 cm long, sulcate, glabrous. Sterile lamina 80–150 cm long, 25–40 cm wide, somewhat reduced below. Fertile lamina 60–140 cm long, 20–40 cm wide, somewhat reduced below. Sterile pinnae up to 32 cm long, 3.9 cm wide, linear, glabrous, petiolate, firmly membranous, acute, acuminate, base unequally cuneate, margin minutely cartilaginous-serrate. Fertile pinnae up to 25 cm long, pinnately divided into narrowly linear segments up to 8 cm long, 0.2 cm wide, adnate to the rhachis or petiolate at the base, glabrous above, completely covered with sporangia below. Fig. 2/1, 2 (p. 7).

UGANDA. Masaka District: Sese Is., *Dawe* 76! (sterile) & Buddu, 1 km N of Bukeri, 12 July 1969, *Lye* 3494! (mildbraedii form); Mengo District: Busiro, *Dawe* 212! (mildbraedii form)

KENYA. Kwale District: Digo, Buda Forest, Nov. 1936, *Dale in F.D.* 3578! & 14.4 km SW of Kwale, Shimba Hills, Pengo Forest, 11 Feb. 1953, *Drummond & Hemsley* 1202! & Ramisi-Mrima Hill road, 7 km after Ramisi R., 23 Mar. 1974, *R.B. & A.J. Faden* 74/309!

TANZANIA. Rufiji District: Mafia I., Dawani District, 9 Aug. 1937, *Greenway* 5036! (mildbraedii form) & 24 Mar. 1933, *Wallace* 760 (tenuifolia form); Masasi District: 35 km E of Masasi, near Ndanda, *Scheven* 50!; Zanzibar: Josani Forest, 25 Nov. 1960, *Faulkner* 2732!

DIST. **U** 4; **K** 7; **T** 1, 6, 8; **Z**; **P**; Bioko, Cameroun, Congo (Kinshasa), Angola, Mozambique, South Africa (Natal & Transkei); Madagascar, Galego I., Comoro Is., and Mauritius

HAB. Terrestrial and epiphytic essentially in swamp forest, and swampy stream bottoms, *Elaeis–Pandanus* associations, mixed evergreen forest, also fringing forest; 0–1200 m

SYN. *Lomaria tenuifolia* Desv. in Mag. Ges. Naturf. Fr. Berl. 5: 326 (1811)

L. meyeriana Kunze in Linnaea 10: 509 (1836). Type: South Africa, Transkei, between the Umtentu and Umzimkulu Rivers, *Drege* (LZ, holo.; BM!, iso.)

Stenochlaena meyeriana (Kunze) C.Presl, Epim. Bot.: 166 (1851) & in Abh. Königl. Böhm. Ges. Wiss. ser. 5, 6: 526 (1851)

Lomariobotrys tenuifolia (Desv.) Fée, Mém. Fam. Foug. 5: 46 (1852)

L. meyeriana (Kunze) Fée , Mém. Fam. Foug.: 46, t. 5, fig. a (1852)

Polybotrya meyeriana (Kunze) Mett., Fil. Hort. Bot. Lips: 24, t. 1, fig. 4, 7 (1856)

Acrostichum meyerianum (Kunze) Hook., Garden Ferns: t. 16 (1862)

Polybotrya tenuifolia (Desv.) Kuhn, Fil. Afr.: 52 (1868)

Acrostichum tenuifolium (Desv.) Bak., Syn. Fil.: 412 (1868)

Lomariopsis tenuifolia (Desv.) Christ., Farnkr.: 42 (1897)

Stenochlaena mildbraedii Brause in E.J. 53: 384 (1915); Tardieu-Blot in Mém. I.F.A.N. 28: 87, t. 39/5, 6 (1953); Alston, Ferns & Fern Allies, W. Trop. Afr.: 50 (1959); Tardieu-Blot, Fl. Cameroun 3: 353, t. 34/5, 6 (1964); Holttum in Amer. Fern J. 61: 121 (1971); Schelpe, C.F.A., Pterid.: 185 (1977). Type: Bioko, above San Carlos, *Mildbraed* 6995 (B, lecto., chosen by Tardieu)

NOTE. I have no doubt that *S. tenuifolia* & *S. mildbraedii* are forms of one species despite the fact that all material from Uganda and Western Africa (very few specimens) has simply pinnate fertile fronds. There are at the two sheets collected by Kirk labelled "Dar Salam Zanzibar" (presumably from the mainland at Dar es Salaam) one pinnate the other bipinnate and a pencilled note (by Baker?) from the same caudex as the other specimen. Both forms occur on Mafia I. and Schelpe mentions that one duplicate of *Junod* 2 from Rikatla is pinnate (K) and another bipinnate (LISC). Field studies are needed; it seems unlikely they occur on the same plant but are minor forms genetically based. *Stenochlaena warneckei* Hieron. in E.J. 46: 382 (1911); F.D.O.-A 1: 83 (1929) is *Lomariopsis warneckei* (Heiron.) Alston.

INDEX TO BLECHNACEAE

No new names are validated in this part

GEOGRAPHICAL DIVISIONS OF THE FLORA

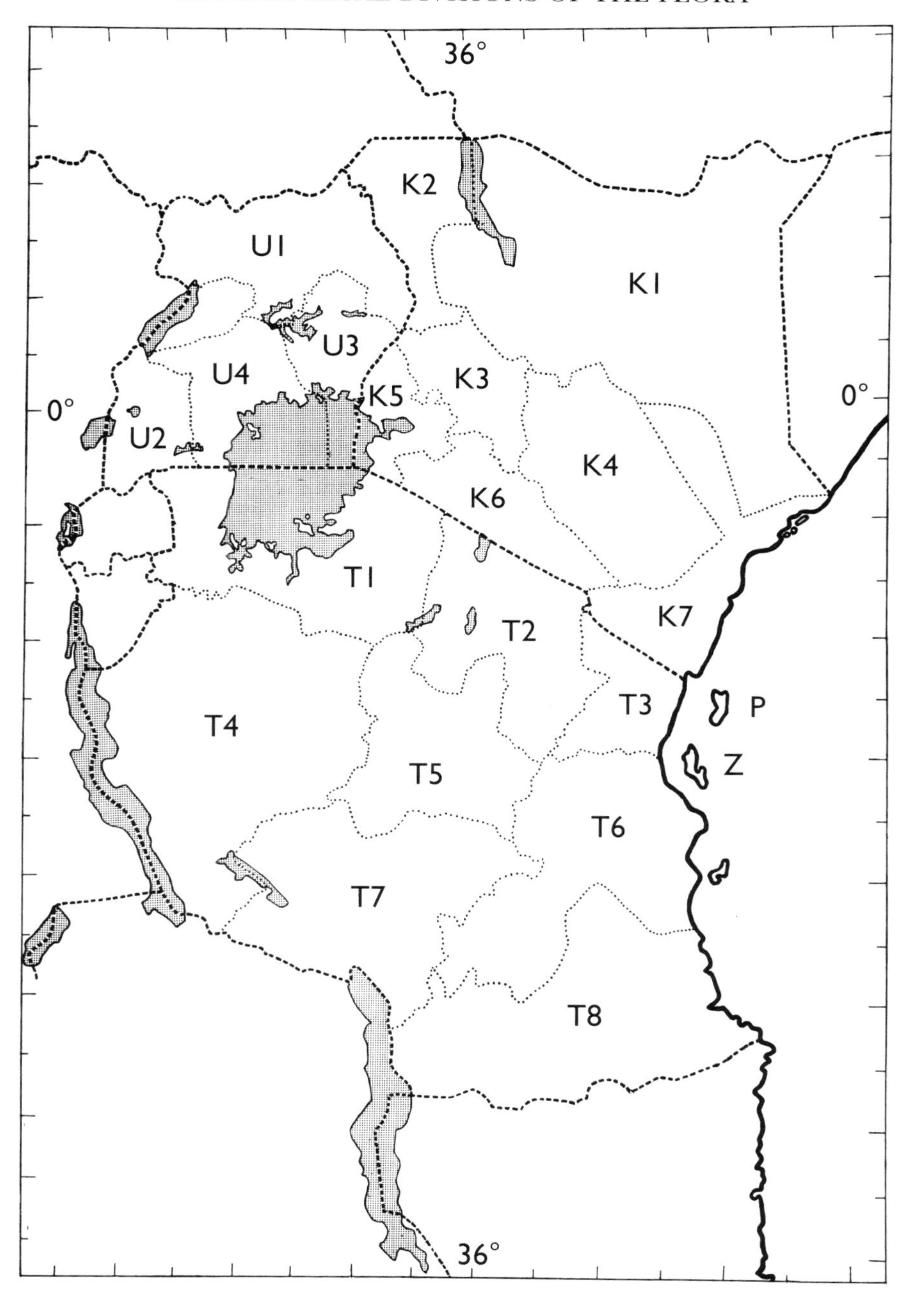

First published in 2006 by
Royal Botanic Gardens, Kew
Richmond, Surrey, TW9 3AB, UK
www.kew.org

ISBN 1 84246 129 X

Design by Media Resources, typesetting and page layout by Margaret Newman,
Information Services Department,
Royal Botanic Gardens, Kew.

Printed in the UK by Hobbs the Printers

For information or to purchase all Kew titles please visit
www.kewbooks.com or email publishing@kew.org